News

Für meinen Ehemann

Alle Rechte in diesem Buch sind der Autorin vorbehalten.

Autorin / Cover / Bilder

Tanja Feiler

Viel ist passiert

Die Cute Pets haben neue Songs geschrieben, ihr Outfitdesign ist ausgestellt worden, die Zeitung hat über die Kuscheltiere berichtet und die Cute Pets spielten am Petcity Theater ein Theaterstück, das sie selbst verfasst haben. Im Zeitungsartikel steht, dass das neue Album der Cute Pets mit Spannung erwartet wird. Die Cute Pets haben ein paar

Coverentwürfe gemacht.

Jetzt haben sie sich für ein Cover entschieden, das ist ihr neues Album.

Cute Pets New Album

1. Daycare - Song
2. Cute Pets Song II
3. Lets go Cute Pets
4. Die niedlichen Kuscheltiere
5. Wir, die WG
6. You rock
7. All time long
8. The new Song
9. Dance with us

Song Lyrics

1. Daycare – Song

   For the city
   That´s so pretty
   The daycare
   For the selfcare

Singing the daycare song all time long!

2. Cute Pets Song II

That's the way
for the day
sing this Song
all time long

3. Lets go Cute Pets

Lets go
for the show
don't forget
we are the Cute Pets
friends and more
and always for
social things

4. Die niedlichen Kuscheltiere

Es ist doch klar
und natürlich wahr
die Kuscheltiere
sind niedlich
und friedlich

5. Wir, die WG

Autoren, Künstler, Designer, Musiker
das sind wir
die WG Cute Pets
Lets
do the best

6.   You   rock

Playing the music
that's a big
Time for all of us
A great mass
You rock
All time long
Sing   the   song

7. All time long

Sing the song
all time long
For the good
thing
When we think
The time is come
All time long

8. The new song

Writing a new song
means working together
better and better
lyrics in the head
made
By the Cute Pets

9. Dance with us

Time for fun

Dancing in the sun

With us around

To the sound

All time long

Playing the song

Besonders Danke ich meinem Ehemann

24

25

www.ingramcontent.com/pod-product-compliance
Lightning Source LLC
Chambersburg PA
CBHW041621180526
45159CB00002BC/965